MAJOR DISASTERS

TEXAS ICE STORMS

BY TRUDY BECKER

WWW.APEXEDITIONS.COM

Apex is distributed by North Star Editions:
sales@northstareditions.com | 888-417-0195

Produced for Apex by Red Line Editorial.

Photographs ©: Shutterstock Images, cover, 9, 10–11, 12–13, 18–19, 22–23, 29; Amanda McCoy/Star-Telegram/AP Images, 1, 6; iStockphoto, 4–5, 14–15, 26–27; Chelsea Purgahn/Tyler Morning Telegraph/AP Images, 7; Connie Jo Discoe/McCook Daily Gazette/AP Images, 8; LM Otero/AP Images, 16–17; Eli Hartman/Odessa American/AP Images, 20, 24; Denise Cathey/The Brownsville Herald/AP Images, 21

Library of Congress Control Number: 2023910166

ISBN
978-1-63738-761-0 (hardcover)
978-1-63738-804-4 (paperback)
978-1-63738-887-7 (ebook pdf)
978-1-63738-847-1 (hosted ebook)

Printed in the United States of America
Mankato, MN
012024

NOTE TO PARENTS AND EDUCATORS

Apex books are designed to build literacy skills in striving readers. Exciting, high-interest content attracts and holds readers' attention. The text is carefully leveled to allow students to achieve success quickly. Additional features, such as bolded glossary words for difficult terms, help build comprehension.

TABLE OF CONTENTS

CHAPTER 1

A STORM ARRIVES 4

CHAPTER 2

WINTER WEATHER 10

CHAPTER 3

ICY DANGER 16

CHAPTER 4

FOR THE FUTURE 22

COMPREHENSION QUESTIONS • 28

GLOSSARY • 30

TO LEARN MORE • 31

ABOUT THE AUTHOR • 31

INDEX • 32

CHAPTER 1

In north Texas, the temperature drops. Clouds form in the sky. Icy rain starts to fall.

Ice and snow are most common in north Texas. The rest of the state tends to stay warmer.

Icy roads make car accidents much more likely.

Soon, the roads are covered in slick ice. It's too dangerous to get to school or work. People stay home. They close their doors tightly against the cold.

In 2018, some roads in Texas became so icy that people used them for sledding.

Suddenly, the power goes out. People huddle in blankets to keep warm. They pull out flashlights and wait for the ice storm to pass.

Ice can build up on power lines and cause them to fall down.

Frozen water pipes can crack and then leak water when they thaw.

WATER PROBLEMS

During some ice storms, water pipes freeze. Other times, buildings that clean water lose power. So, water in houses might not be safe. People must boil the water before using it.

CHAPTER 2

WINTER WEATHER

An ice storm is a type of winter storm. It happens when freezing rain falls and ice builds up. These storms sometimes happen in Texas.

In Texas, ice and sleet are more common than snow.

Texas winters are usually mild. In Texas, only a few days of freezing rain each year are expected. But in the 2010s, stronger ice storms became more common.

STORMS OF THE PAST

Texas has had many winter storms. In 1887, a storm knocked down trees and **telegraph** wires. In 1929, some places got more than 20 inches (51 cm) of snow.

Freezing rain turns to ice when it hits things. Thick ice can make tree branches break and fall.

Winter Storm Uri made cities throughout Texas freezing cold. Many places lost power.

The 2020s saw several very strong storms. Winter Storm Uri hit in 2021. And Winter Storm Mara came in 2023. The long, cold storms damaged buildings. They harmed people, too.

FAST FACT

The 2021 ice storm was nicknamed Snowmageddon.

CHAPTER 3

ICY DANGER

When ice storms hit, **black ice** covers the pavement. Cars crash. Drivers slide off the roads. People try to stay home until it is safe.

If ice coats airport runways, planes can't take off or land. Flights are delayed or canceled.

Without power, refrigerators and freezers can't work. Stores may run low on food.

Heavy ice drags down power lines. Millions of people can lose power. Without power, people can't keep their homes warm. They may also struggle to get or store food.

FAST FACT

During the 2021 ice storm, Texas's **power grid** nearly failed.

Ice storms can be deadly, too. In the 2021 storm, around 250 people died. Some got **hypothermia**. Others were hurt in crashes.

Cities set up warming shelters where people without heat could stay.

To save cold-stunned turtles, people move them to warmer areas.

WILDLIFE

Ice storms affect animals, too. If water gets too cold, fish will freeze. Sea turtles will become cold-stunned. They grow weak and slow. If they don't warm up, they will die.

CHAPTER 4

FOR THE FUTURE

Events like the Texas ice storms are linked to **climate change**. Warm places like Texas can face stronger winter weather than before.

To prepare for more storms, some cities plan to get more snowplows and trucks that spread salt or sand.

3 RD ST
600 E
STOP

Texas leaders wanted to handle future storms better. After the 2021 storm, the government fixed the power grid. In 2023, leaders tried to improve warning systems.

A WIDE WARNING

Scientists send out winter storm warnings. They alert people when dangerous weather is coming. During the 2021 storm, alerts went to all 254 Texas counties.

Loss of power was one of the biggest dangers during the 2021 storm.

Some homes use batteries as a backup. They can provide power if there is an outage.

Burying power lines could help as well. So could sending out more **emergency response teams**. These actions would be very expensive. But they could help keep more Texans safe.

COMPREHENSION QUESTIONS

Write your answers on a separate piece of paper.

1. Write a few sentences explaining the main ideas of Chapter 4.

2. Would you rather live in a very warm place or a very cold place? Why?

3. Which strong ice storm hit Texas in 2023?

   A. Winter Storm Mara
   B. Winter Storm Uri
   C. Snowmageddon

4. How would fixing the power grid help people in future storms?

   A. People would be less likely to lose their heating.
   B. Roads and runways would be safer to use.
   C. Cities would not get as much ice.

5. What does **mild** mean in this book?

*Texas winters are usually **mild**. In Texas, only a few days of freezing rain each year are expected.*

A. very cold and rainy
B. not very cold or rainy
C. not very spicy

6. What does **alert** mean in this book?

*Scientists send out winter storm warnings. They **alert** people when dangerous weather is coming.*

A. let people know something
B. don't tell people something
C. hide something from people

Answer key on page 32.

GLOSSARY

black ice

A thin, see-through layer of ice that forms on dark surfaces, such as roads.

climate change

A dangerous long-term change in Earth's temperature and weather patterns.

counties

Areas within a state.

emergency response teams

Groups of people who go to areas after storms or other disasters and work to help.

hypothermia

When a person's body temperature drops dangerously low.

power grid

The network of power lines that brings electricity to people.

telegraph

A way of sending messages over wires that people used in the past.

BOOKS

Abdo, Kenny. *How to Survive a Blizzard*. Minneapolis: Abdo Publishing, 2019.

Dalgleish, Sharon. *Blizzards*. Mendota Heights, MN: Apex Editions, 2023.

Pettiford, Rebecca. *Ice Storms*. Minneapolis: Bellwether Media, 2020.

ONLINE RESOURCES

Visit **www.apexeditions.com** to find links and resources related to this title.

ABOUT THE AUTHOR

Trudy Becker lives in Minneapolis, Minnesota. She loves visiting her nieces in Texas (when there are no storms).

INDEX

A
animals, 21

B
black ice, 16

C
climate change, 22

E
emergency response teams, 27

H
hypothermia, 20

P
pipes, 9
power grid, 19, 25
power lines, 19, 27

R
rain, 4, 10, 12
roads, 6–7, 16

W
water, 9, 21
Winter Storm Mara, 15
Winter Storm Uri, 15
winter storm warnings, 25

ANSWER KEY:
1. Answers will vary; 2. Answers will vary; 3. A; 4. A; 5. B; 6. A